Exploration of the Gas Giants, Ice Giants, and Minor Planets.

Patrick H. Stakem

(c) November 2022

Table of Contents

Introduction..3
    Mars helicopter scout 3
    Comets..5
    Jupiter...7
    Flyby mission - Pioneer program (1973 and 1974) 9
    Voyager program (1979)...10
    Ganymede Exploration...10
    Europa Exploration 10
    Callisto Exploration 11
    Io 11
    Saturn 12
    Pioneer 11 flyby 13
    Voyager...13
    Cassini orbiter 14
Uranus, Neptune, and Pluto 16
    Pluto 16
    Further Out 17
    The Sun 18
    Solar space missions 18
Glossary 23
References..25
Resources 29
If you enjoyed this book, you might be interested in some of his others as well. 31

Introduction

Since the 1990s, a total of 13 minor planets – currently asteroids or dwarf planets – have been visited by space probes. A minor planet is in direct orbit around the sun, but is not a planet or comet.

The International Astronomical Union defines a planet as an object in orbit around the Sun, with sufficient mass to achieve hydrostatic equilibrium (it's round), and it has cleaned up its orbit.

In addition to the listed objects, have been imaged by spacecraft at distances too large to resolve features (over 100,000 km), and are hence not considered as "visited". Asteroid 132524 APL was imaged by New Horizons in 2006 at a distance of 101,867 km, 2685 Masursky by Cassini in 2000 at a distance of 1,600,000 km, and 307 Nike by Pioneer 10 in 1972 at a distance of 8,800,000 km. The Hubble Space Telescope, a spacecraft in Earth orbit, has imaged several large asteroids, including 2-Pallas and 3-Juno.

A successor mission to the existing Mars Helicopter Scout is in the planning stage, the Mars Science Helicopter. In addition, the Mars rovers will be discontinued, and helicopters will be used to retrieve samples for the Mars Sample Return Mission. It will be called Fetch. Mars has water today, mostly vapor in the atmosphere, and brine's in the soil. Sorry, no canals. No lakes or seas. The north polar ice cab has visible water ice. There is also water ice under the frozen carbon dioxide at the southern pole. Water on the surface of Mars would sublime (go from solid to vapor) into space. It's possible that in the past, Mars had about 1/3 of its surface covered in water ice. Aeolis Palus in Gale Crater, explored by the Curiosity rover, is the geological remains of an ancient freshwater lake that could have been a hospitable environment for microbial life. Mars lacks a thick atmosphere, ozone layer, and magnetic field, allowing solar and cosmic radiation to strike the surface unimpeded.

There is evidence of several subglacial lakes on Mars, 1.5 km (0.93 mi) below the southern polar ice cap. Water vapor was not detected on Mars until 1963.

Why go all the way there when you can get them delivered? Over 60 meteorites have been found that came from Mars. Mars meteorites usually are found in Antarctia. My old college professor at Carnegie Mellon figured out why, but I can't follow the math. He had 5 degrees beyond the Bachelors.

The Mariner-9 spacecraft sent images of large river valleys, broken rock dams, and mapped 40,000 river valleys. There are also the outlines of lake basins. In 2012, NASA announced that the Curiosity rover found direct evidence for an ancient stream bed in Gale Crater.

Eridania Lake is a theorized Martian ancient lake with a surface area of roughly 1.1 million square kilometers. Its maximum depth was 2,400 meters and its volume was 562,000 km3. It was larger than the largest landlocked sea on Earth, the Caspian and held more than nine times as much water as all of North America's Great Lakes combined. There were also river deltas on Mars, over a wide range. the InSight lander uncovered unexplained magnetic pulses, and oscillations consistent with a planet-wide reservoir of liquid water deep underground.

The existence of ice in the Martian northern (Planum Boreum) and southern (Planum Australe) polar caps has been known since the time of Mariner 9 orbiter.

Another mission for a helicopter is the surface of Titan, with a mission called Dragonfly. Titan has a carbon rich chemistry and could support extraterrestrial habitability. The mission will launch in 2027. It has 8 sets of rotors, and uses a RTG for power. It can travel up to 22 mph, and can climb up to 4 kilometers. The dense atmosphere of Titan assists the helicopter. It will be able to travel for several kilometers. It has lithium-ion battery's, recharged from the RTG. The entire craft weights about 450kg. Titan has the thickest atmosphere of any moon in our solar system.

The instrument cluster includes a mass spectrometer, a gamma ray and neutron spectrometer, a suite of meteorology sensors, and a microscope and cameras.

Currently, Dragonfly is scheduled to launch in 2027. The communication with Earth takes 70-90 minutes.

Comets

Debate continues about how much water ice is in a comet. In 2001, the Deep Space 1 spacecraft obtained high-resolution images of the surface of Comet Borrelly. It was found that the surface of comet Borrelly is hot and dry, with a temperature of between 26 to 71°C (79 to 160 °F), and extremely dark, suggesting that the ice has been removed by solar heating and maturation, or is hidden by the soot-like material that covers Borrelly's. In July 2005, the Deep Impact probe blasted a crater on Comet Tempel-1 to study its interior. The mission yielded results suggesting that the majority of a comet's water ice is below the surface and that these reservoirs feed the jets of vaporized water that form the coma of Tempel 1. The mission was renamed EPOXI, and it made a flyby of Comet Hartley 2 on November 4, 2010.

Data from the Stardust mission show that materials retrieved from the tail of Wild 2 were crystalline and could only have been "born in fire," at extremely high temperatures of over 1,000 °C (1,830 °F). Although comets formed in the outer Solar System, radial mixing of material during the early formation of the Solar System is thought to have redistributed material throughout the proto-planetary disk, so comets also contain crystalline grains that formed in the hot inner Solar System. This is seen in comet spectra as well as in sample return missions. More recent still, the materials retrieved demonstrate that the "comet dust resembles asteroid materials". These new results have forced scientists to rethink the nature of comets and their distinction from asteroids.

Starting in 1801, astronomers discovered Ceres and other bodies between Mars and Jupiter that for decades were considered to be planets. Between then and around 1851, when the number of planets had reached 23, astronomers started using the word asteroid for the smaller bodies and began to distinguish them as minor planets rather than major planets (Courtesy, Wikipedia).

ESA's Rosetta probe is presently in erratic orbit around Comet Churyumov–Gerasimenko. On November 12, 2014, its lander Philae successfully landed on the comet's surface, the first time a spacecraft has ever landed on such an object in history. It is currently in hibernation. Last contact was in November 2014.

The Uranian system has been studied up close only once, by the spacecraft Voyager 2 in January 1986. It took several images of the moon Umbriel. We need more data.

Jupiter

The exploration of Jupiter has been conducted via close observations by automated spacecraft. It began with the arrival of Pioneer 10 into the Jovian system in 1973, and, as of 2014, has continued with seven further spacecraft missions. All of these missions were undertaken by the National Aeronautics and Space Administration (NASA), and all but one have been flybys that take detailed observations without the probe landing or entering orbit. These probes make Jupiter the most visited of the Solar System's outer planets as all missions to the outer Solar System have used Jupiter flybys to reduce fuel requirements and travel time. Plans for more missions to the Jovian system are under development, none of which are scheduled to arrive at the planet before 2016. Sending a craft to Jupiter entails many technical difficulties, especially due to the probes' large fuel requirements and the effects of the planet's harsh radiation environment.

The first spacecraft to visit Jupiter was Pioneer 10 in 1973, followed a year later by Pioneer-11. Aside from taking the first close-up pictures of the planet, the probes discovered its magnetosphere and its largely fluid interior. The Voyager-1 and Voyager-2 probes visited the planet in 1979, and studied its moons and the ring system, discovering the volcanic activity of Io and the presence of water ice on the surface of Europa. Ulysses further studied Jupiter's magnetosphere in 1992 and then again in 2000. The Cassini probe approached the planet in 2000 and took very detailed images of its atmosphere. The New Horizons spacecraft passed by Jupiter in 2007 and made improved measurements of its and its satellites' parameters.

The Galileo spacecraft is the only one to have entered orbit around Jupiter, arriving in 1995 and studying the planet until 2003. During this period Galileo gathered a large amount of information about the Jovian system, making close approaches to all of the four large Galilean moons and finding evidence for thin atmospheres on three of them, as well as the possibility of liquid water beneath their surfaces. It also discovered a magnetic field around Juipter

As it approached Jupiter, it also witnessed the impact of Comet Shoemaker–Levy 9. In December 1995, it sent an atmospheric probe into the Jovian atmosphere, so far the only craft to do so.

Future probes by NASA included the Juno spacecraft, launched in 2011, which will enter a polar orbit around Jupiter to determine whether it has a rocky core, among other objectives. The European Space Agency selected the L1-class JUICE mission in 2012 as part of its Cosmic Vision program to explore three of Jupiter's Galilean moons, with a possible Ganymede lander provided by Roscosmos. JUICE is proposed to be launched in 2022, with a projected 3.5 year working life. What do they hope to learn at Ganymede?

- Characterization of the ocean layers and detection of subsurface water reservoirs.
- Topographical, geological, and compositional mapping of the surface.
- The physical properties of the icy crusts.
- Characterization of the internal mass distribution, dynamics and evolution of the interiors.
- Investigation of Ganymede's tenuous atmosphere.
- Nature of Ganymede's intrinsic magnetic field and its interactions with the Jovian magnetosphere.

Flyby mission - Pioneer program (1973 and 1974)

These missions were the work of NASA-Ames, comprising a space weather network. Pioneer-6 (A) was launched in December of 1965, followed by Pioneer -7 (B) in 1966. Pioneer-8 (C) launched in 1967. Pioneer-9 (D) launched in 1968. It has been silent since 1983). Pioneed-E was lost to a launch failure in August 1969. Pioneer 6 and 9 are in a 1.1 AU orbit. There can be in an orbital position that cannot be seen from Earth. It gets the data several days before ground based observatory's.

Pioneer 10 was a Jupiter Interstellar mission. Pioneer 11 was a Jupiter-Saturn interstellar medium mission. Pioneer-H was never launched.

Voyager program (1979)

The Pioneer Venus Orbital and Multi-probes launched in 1978.

Ulysses (1992)
Cassini (2000)
New Horizons (2007)

Orbiter missions
Galileo (1995–2003)
Juno (2016)
Jupiter Icy Moon Explorer (Juice) (ESA)(2023)

The Pioneer Venus probe, Large Probe, North Probe, and Day Probe were launched in 1978.

Ganymede Exploration

Ganymede is a moon of Jupiter, the largest and most massive of moons in our solar system. It doesn't have a substantial atmosphere, as well as a magnetic core, and no atmosphere. It is composed of silica rock, and water. It has a magnetic field, due to convection of the liquid iron core. Its atmosphere has oxygen. We do not know about its ionosphere.

Europa Exploration

Five spacecraft have closely visited Europa, and it is under the watchful eye of the Hubble Space Telescope. It has been observed by Pioneer 10 and 11, Voyager 1 and 2, and Galileo. Europa was the first moon discovered beyond Earth. Planned missions are NASA's Europa Clipper (2024) and ESA's 2023 JUICE mission.

Callisto Exploration

Callisto is the second largest moon of Jupiter, and the third largest in our solar system. It was discovered by Galileo. It is about the same size of the planet Mercury. It mostly consists of rocks and ices. Carbon dioxide and organic compounds have been observed on the surface. The surface is heavily cratered. It has a thin atmosphere of carbon dioxide. Callisto has been considered as the best place for a human base for exploration of the Jovian system.

Io

Whereas the Earth's magnetosphere is shaped by the solar wind, Jupiter's is shaped by it's moon, Io's plasma. The Jovian magnetosphere is torus shaped, and rotates with the planet. The magnetosphere is a particle accelerator, using the intense magnetic field to drive the particles to high velocities. The magnetic field itself is caused by the rotation of Jupiter's core of liquid metallic hydrogen. Volcanic activity on the moon Io releases compounds of sulphur. These are ionized by solar ultraviolet radiation, and form into a plasma torus, and ring current around Jupiter This in turn interacts with the Jovian magnetic field. Io is one of the Galilean moons. It has over 400 active volcanoes, and it's surface is comprised of planes of sulphur and sulphur dioxide. Along with the other Galilean satellites, Io is responsible for the adoption of the Copernican model of the solar system, Kepler's laws, and the first measurement of the speed of light.

Saturn

The exploration of Saturn has been solely performed by unmanned probes. Three missions were flybys, which formed an extended foundation of knowledge about the system. The Cassini–Huygens spacecraft was launched in 1997 and Cassini is currently in orbit (as of 2015).

Saturn is the only planet in our solar system less dense than water. It's outer atmosphere is mostly hydrogen, with some helium. There are traces of ammonia, acetylene, ethane, propane, phosphine, and methane. The winds on Saturn are the second fastest among the Solar System's planets, after Neptune's. Hubble Space Telescope noticed a persistent hexagonal pattern at the polar vortex, and there is a jet stream. The polar vortex, the size of the Earth, has been around possibly for millions of years. Saturn has a simple magnetic bipole.

Saturn has no known Trojans. It does have 83 known moons. Trojans are asteroids in the same orbit. Huygens saw the rings in 1655. Most known Trojans share the orbit of Jupiter around the Sun. There may be more than a million Jupiter trojans. Mars has nine, Earth has two.

Pioneer 11 Satrun flyby

This mission made the first fly-by of Saturn in 1979. In 1980, Voyager-1 visited the Saturnian system. About a year later, Voyager-2 returned more data, and they went on to Uranus.

Saturn was first visited by Pioneer 11 in September 1979. It flew within 20,000 km of the top of the planet's cloud layer. Low-resolution images were acquired of the planet and a few of its moons; the resolution of the images was not good enough to discern surface features. The spacecraft also studied the rings; among the discoveries were the thin F-ring and the fact that dark gaps in the rings are bright when viewed towards the Sun, or in other words, they are not empty of material. Pioneer 11 also measured the temperature of Titan at 250 K.

Voyager

In November 1980, the Voyager 1 probe visited the Saturn system. It sent back the first high-resolution images of the planet, rings, and satellites. Surface features of various moons were seen for the first time. Because of the earlier discovery of a thick atmosphere on Titan, the Voyager controllers at the Jet Propulsion Laboratory elected for Voyager 1 to make a close approach of Titan. This greatly increased knowledge of the atmosphere of the moon, but also proved that Titan's atmosphere is impenetrable in visible wavelengths, so no surface details were seen. The flyby also changed the spacecraft's trajectory out from the plane of the Solar System which prevented Voyager 1 from completing the Planetary Grand Tour of Uranus, Neptune & Pluto.

Almost a year later, in August 1981, Voyager 2 continued the study of the Saturn system. More close-up images of Saturn's moons were acquired, as well as evidence of changes in the rings. Voyager 2 probed Saturn's upper atmosphere with its radar, to measure temperature and density profiles. Voyager 2 found that at the highest levels (7 kilopascals pressure) Saturn's temperature was 70 K (−203°C) (i.e. 70 degrees above absolute zero), while at the deepest levels measured (120 kilopascals) the temperature increased to 143 K (−130°C). The north pole was found to be 10 K cooler, although this may be seasonal. Unfortunately, during the flyby, the probe's camera platform stuck for a couple of days and some planned imaging was lost. Saturn's gravity was used to direct the spacecraft's trajectory towards Uranus.

The probes discovered and confirmed several new satellites orbiting near or within the planet's rings. They also discovered the small Maxwell and Keeler gaps in the rings.

Cassini orbiter

On July 1, 2004, the Cassini–Huygens spacecraft performed the SOI (Saturn Orbit Insertion) maneuver and entered into orbit around Saturn. Before the SOI, Cassini had already studied the system extensively. In June 2004, it had conducted a close flyby of Phoebe, sending back high-resolution images and data.

The orbiter completed two Titan flybys before releasing the Huygens probe on December 25, 2004. Huygens descended onto the surface of Titan on January 14, 2005, sending a flood of data during the atmospheric descent and after the landing. During 2005 Cassini conducted multiple flybys of Titan and icy satellites.

On March 10, 2006, NASA reported that the Cassini probe found evidence of liquid water reservoirs that erupt in geysers on Saturn's moon Enceladus.

On September 20, 2006, a Cassini probe photograph revealed a previously undiscovered planetary ring, outside the brighter main rings of Saturn and inside the G and E rings.

In July 2006, Cassini saw the first proof of hydrocarbon lakes near Titan's north pole, which was confirmed in January 2007. In March 2007, additional images near Titan's north pole discovered hydrocarbon "seas", the largest of which is almost the size of the Caspian Sea.

As of 2009 the probe has discovered and confirmed four new satellites. Its primary mission ended in 2008 when the spacecraft completed 74 orbits around the planet. In 2010, the probe began its first extended mission, the Cassini Equinox Mission. It is now currently on its second mission extension, the Cassini Solstice Mission, expected to last through September 2017.

Uranus, Neptune, and Pluto

Uranus is the seventh planet out from the Sun. It and Neptune are considered as Ice Giants. These planets lack a well-defined solid surface. Their atmosphere is similar to Jupiter and Saturn's as Hydrogen and helium, with more water, ammonia, and methane ices. The ice giants lack a solid surface, but have a complex layered cloud structure. Uranus has a ring system, a magnetosphere, amd numerous moons. The planets axis of rotation is almost 90 degrees. Voyager 2 is the only spacecraft to visit Uranus. Observations from Earth show seasonal change in wind speeds exceed 560 mph. Sir William

Herschel first observed Uranus on 13 March 1781. He thought it was a comet.

Uranus orbits the Sun every 84 years. It rotates every 17 hiurs, 14 minutes, causing strong winds. It has the mass of 14.5 times Earth's. It has a water-ammonia ocean. In 2020, NASA noted a large atmospheric magnetic bubble, as well as 2 known trojans.

Extraterrestrial diamonds are probably common on the icy giants.

Pluto

Pluto, our furthest-out planet has two moons. It is a dwarf planet in the Kuiper belt. It is the nineth largest object to orbit the Sun. It consists of ice and rock. It is 30-49 AU from the Sun. It has four known moone. It does venture within the orbit of Neptune, but there has, to date, been no collision. It has 5 known moons, where Charon is half the size of its primary. The New Herizons mission visited Pluto, in 2015. In 2016m, the International Astronomical Union changed the definition of planet, and "Pluto" was excluded the cut. We still believe in you, Pluto.

The Kuiper belt extends from the orbit of Neptune at 30 AU to 50 AU. It is similar to the asteroid belt. Most objects in the Kuiper belt objects are not solids, but frozen volatile's. Besides Pluto, the belt is the home to the dwarf planets Orcus, Haumea, Quaoar, and Makemake, There are thousands of Kuiper belt objects.

Further Out

In 2014, astronomers hypothesized the existence of further planets beyond Neptune, the tenth planet. There were anomaly's that pointed to another planets, and Percival Lowell was on the hunt for Planet-X. This was to explain the apparent discrepancy's in the orbits of Uranus and Neptune. Clyde Tombough's discovery of Pluto in 1930 appeared to validate the hypothesis. It was not big enough to explain the discrepancy;s, and a tenth planet was speculated. Although Pluto was named a planet in 1930, but it was to small to produce the observed effects, and a tenth planet was no found. The search was not abandoned until the 1990's when the data from the satellite explorer Voyager-2 showed the irregularity's in Uranus" orbit were due to an error in investigation of Neptune's mass. Along the way, numerous small icy objects were discovered and characterized as dwarf planets. The International Astronomical Union declared these objects, including Pluto, as dwarf planets. A new object, named Planet nine, was found to be a super-earth or gas giant.

In 1894, Percival Lowell founded the Lowell Observatory in Arizona. Not finding anything that matched what he was looking for, he continued the search in 1914-16. He searched along the ecliptic plane, which housed orbits of known planets.

Gliese 1214b

This ocean planet was nicknamed "the water world." It is a super-Earth in size, and orbits the sun Gilese, 48 light-years from our Sun. It was discovered in 2009. It transits its parent star, which allows for spectroscopic examination of the planet's atmosphere. NASA announced in 2013 that clouds may have been detected in the planet's atmosphere.

What can we learn about these very distant planets, and how can we do it? The radius of GJ 1214 b can be inferred from the amount of dimming when the planet crosses in front of its parent star as viewed from Earth. The mass of the planet can be inferred from sensitive observations of the parent star's radial velocity, measured through small shifts in stellar spectral lines due to the Doppler effect. Knowing the planet's mass and radius, its density can be calculated. Through a comparison with theoretical models, the density provides limited but highly useful information about the composition and structure of the planet. Can we detect life on the subject planet? Not yet. If it is a waterworld, it could possibly be thought of as a bigger and hotter version of Jupiter's Galilean moon Europa.

What might give us a clue that the planet is inhabited? We might notice additional satellite's appearing around the target planet. These would be detected by their transits. There's not much to see.

JWST
The James Webb Space Telescope is the follow-on to the Hubble Space Telescope. It uses updated technology and a new approach for the mirror, using 18 hexagonal segments, that are individually adjustable. There are 126 small motors to adjust the optics for fine tuning. Ball Aerospace was the principal optical contractor. The resulting mirror is 6.5 meters in diameter, compared to Hubble's 2.4 meter. . JWST observes in long wavelengths visible through the mid-infrared. The spacecraft was placed in a halo orbit at the Sun-Earth L2 Lagrange point about a million miles from Earth. It has a large sun shield to block the Sun's light from interfering with the observations. The project was the top pick in NASA's 2000 Decadal Survey. Work has been going on since 1989, primarily at the Goddard Space Flight Center, the lead center for the project. JWST observes in the infrared, where Hubble observes in the visible and infrared. Infrared goes through dust clouds in space better than the visible.

Primary Mirror

The primary mirror has 18 hexagonal segments, that can be individually adjusted. They are made of beryllium coated with gold. The mirror is 6.5 meters in diameter. It has a light-collecting surface of 25 square meters. Some of the mirror is obstructed by the support struts. Since JWST observes in the infrared, it is necessary to keep the mirror very cold, below 50K. Otherwise you get infrared "noise" from the spacecraft itself. Webb can also observe planets in our solr system, and their moons, as well as the Kuiper belt objects.

Two is better than one

Since HST is still operational when JWST went to orbit, the two can be used together. The ESA telescope Herschel was in the vicinity of L2 when JWST arrived.

Next Telescope

Even though JWST is the new kid on the block, planning has been underway for its successor, the Wide Field Infrared Survey (telescope), sometimes call the Super Hubble.

The Wfirst / Nancy Grace Roman Space Telescope is in development. It was originally called the "Wide-Field Infrared Survey Telescope.

Wfirst was renamed Nancy Grace Roman after a notable American Astronomer who served as NASA's Chief of Astronomy. It will be launched by 2027. It will have a 2.4 meter primary mirror that will be used by the Wide Field Instrument and a near-infrared camera. In the 2030's, the new 'scope will be looking at some of the estimated 400 billion stars in the Milky Way. It will have the ability to block the light from the parent star, to better study the exo-planets. It will be looking for Earth -size planets within the habitable zone. It will study the planet's atmosphere to check for transformation of the atmosphere by life. This goes far beyond what we can do today, even with JWST, where exoplanets are detected by their gravitation effects on their "Sun."

With what we have to work with at the moment, viewing a star, any Earth sized planet in an Earth like orbit could not be discerned. We currently know of more than 5,000 exoplanets. When sunlight from a star passes through a planets atmosphere, there is both an emissions and a absorption process. These can show us presence or absence of certain gases and compounds. This is called transit spectroscopy.

Roman will also orbit the same libration point that JWST is orbiting, but the orbit will be selected so no collision is possible.

Afterword

Glossary

AAS – American Astronomical Society.
Albedo – reflectivity.
AO – adaptive optics.
APL – Applied Physics Lab of the Johns Hopkins University.
ASIN – Amazon Standard Inventory Number.
Asteroseismology – study of oscillations in stars.
Astro-oceanography - science of extraterrestrial oceans.
Athena - Advanced Telescope for High Energy Astrophysics, ESA Cosmic vision Program.
AU – Astronomical Unit, the distance from Earth to the Sun equal to about 150 million kilometers (93 million miles) or about 8 light minutes.
Big Bang – current cosmological model for the Universe.
Binary star – two stars in orbit around a common point.
Black hole – a collapsed star, compressed so dense that not even light can escape; a singularity.
Blazar – active galactic nucleus with a relativistic jet.
CCD – Charge Coupled Device (like in your cell phone camera)
Centaur – a minor planet in an unstable orbit, behaving like an asteroid or comet.
CGRO – Compton Gamma Ray Observatory
Cheops – ESA, Characterizing ExOPlanets Satellite
Chemolithotrophy, the growth of organisms based on non-organic reduced compounds.
Cluster – groups of stars.
COBE - Cosmic Background Explorer.

Cosmic ray – high energy radiation, from outside the solar system.
Cyro-volcanism – an ice volcano, spewing water, methane, ammonia, and other liquid volatiles.
CXO – Compton X-ray Observatory.
Dark Matter – existence postulated. Might account for 85% of the matter in the known universe.
Dwarf planet - small object in orbit around the Sun. Pluto is an example.

Dwarf star – small star, much smaller than our Sun. Comes in white, red, blue and black variations.
EBL – Extragalactic background light.
EGRET - Energetic Gamma Ray Experiment Telescope (CGRO).
EIS - Europa Imaging System.
ESA – European Space Agency.
E-THEMIS - Europa Thermal Emission Imaging System.
Europa-UVS - Europa Ultraviolet Spectrograph.
EUV – extreme ultraviolet, wavelengths from 10nm to 124nm.
ev – electron volt, unit of energy.
Exa- 10^{19}
FGST – Fermi Gamma ray Space Telescope.
FUSE - Far Ultraviolet Spectroscopic Explorer.
Gas giant – A large planet consisting mostly of hydrogen and helium.
GRB – gamma ray burst
GRO – Gamma Ray Observatory.
GSFC – Goddard Space Flight Center
HEFT – High Energy Focusing Telescope.

HEIM - Habitable Exoplanet Imaging Mission.
Hill Sphere - the region where a planet's gravity dominates that of its star so it can hold on to its moons.
HST – Hubble Space Telescope.
Hubble – Space Telescope named after Edwin Hubble.
Hubble Constant – rate of expansion of the universe.
IAU – International Astronomical Union.
Ice-1I - a different form of water ice, with a different crystalline structure. There are also Ice-II, III, IV, and -V.
Ice giant – A large icy/liquid planet, consisting of elements heavier than hydrogen and helium.
ICEMAG - Interior Characterization of Europa using Magnetometry, on Europa Clipper.
IRAS – Infrared Astronomical Satellite.
ISBN – International Standard Book Number.
JAXA – Japan Space Agency.
JDEM – Joint Dark Energy Mission (NASA, DOE)
JGO – (ESA) Jupiter Ganymede Orbiter.
Jovian – pertaining to Jupiter.
JPL – Jet Propulsion Lab, operated by Cal Tech for NASA; responsible for all deep space missions.
JUICE – ESA Mission to Io, Europa, Callisto, and Ganymede.
KAO – (NASA) Kuiper Airborne Observatory.
LASS - Large-Area Sky Survey, HEAO-1.
LGM – little green men.
Light pollution – interference from background sources.
LWS – Living with a Star, NASA Program.

Maspex - Mass Spectrometer for Planetary Exploration, Europa Clipper
Microlensing – bending of light by massive objects.
MISE - Mapping Imaging Spectrometer for Europa.
Moon – object in orbit around a planet.
Nebula – interstellar cloud of dust and gasses.
NeN – (NASA) near Earth network
NEO – near Earth object.
Neutron star – collapsed core of a large star. Very dense.
NGC – New General Catalog (of Nebulae and Clusters of Stars).
Nova – transient astronomical event involving a bright new star that fades over time.
NSSDC - (U.S.) National Space Science Data Center.
OSEAN – NASA Ocean Worlds Exploration Program.
Parsec – parallel second of arc, unit of length, about 3.26 light years.
Planet – object in orbit around a star.
Planetary disk – debris disks around a star.
Planetoid - asteroid, minor planet, orbit around the Sun.
Perijove – the point in an orbit closest to Jupiter by an object that orbits it.
Photosphere - luminous envelope of a star.
PIMS - Plasma Instrument for Magnetic Sounding, Europa Clipper.
Pulsar – highly magnetized rotating neutron star of a white dwarf.
REASON - Radar for Europa Assessment and Sounding: Ocean to Near-surface, Europa Clipper.
Quark – an elementary particle.
Quasar – quasi-stellar object, galactic nucleus.

Red Shift – an apparent shift of electromagnetic radiation toward an increasing wavelength due to the doppler effect.
Roche's limit – withing 2.44 radii of the planet, no stable moon is possible, due to tidal forces from the primary.
Rogue planet – planet not associated with a star.
RTG - radio-isotope thermoelectric generator.
SALSA - Subglacial Antarctic Lakes Scientific Access.
SAO - Smithsonian Astrophysical Observatory.
SD – scattered disk – contains trans-Neptunian objects.
SERC – Science and Engineering Research Council (U.K.)
SETI – search for extra-terrestial intelligence.
SOFIA – NASA Stratosphere Observatory for Infrared Astronomy.
Solar System – A star and its associated planets and such.
Strangelet – a hypothetical particle, made of up, down, and strange quarks.
SUDA - Surface Dust Analyzer, Europa Clipper
TandEM - Titan and Enceladus Mission.
TESS - Transiting Exoplanet Survey Satellite.
Thalassogen – substance capable of forming a planetary ocean. May not be able to host life. Attributed to Isaac Asimov.
TNO – Trans-Neptunian objects.
Transit Spectroscopy – analyzing star light through a planets atmosphere.
Trojan - minor planet that shares an orbit with one of the larger planets.
TSSM- Titan Saturn System Mission.

UV – ultraviolet, 19 nm to 400 nm wavelength.
White dwarf – very dense remnant of a stellar core.
WISE X-ray - 0.1 to 10 nanometer wavelength.
X-ray binary (star) – binary star, emitting x-rays.- Wide Field Infrared Survey Explorer.
YSO – young stellar objects.
Zombie-sat – dead satellite, in orbit.

References

Asimov, Isaac, *The Left Hand of the Electron,* (essay "The Thalassogens") 1972, ISBN- 978-0856173783.

Associated Press, *The Hubble Space Telescope: A Universe of New Discovery*, 2015, ISBN-1633530469.

Berg, Lee *Off-Earth Refueling: Water, Water, Nearly Everywhere,* 2018, ASIN- B07J5LSSY6.

Carroll Michael, *Ice Worlds of the Solar System: Their Tortured Landscapes and Biological Potential,* 2019, N-ASIN-B07ZQN2Y95.

Cerullo, Mary M. *Antarctic Life Under Ice* 2nd edition; *Exploring Antarctic Seas,* 2019, ISBN-978-0884487470.

Copernicus, Nicolaus *On the Revolutions of the Heavenly Spheres,* 1543, translated from Latin, ASIN-B01MS8TGOV.

Cornish, Neil J. "The Lagrange Points, " 1998. https://math.ucr.edu/home/baez/lagrange.html

Cousteau, Jacques-; Schepp, Steven *Jacques Cousteau: The Ocean World,* 1985, ISBN-978-0810980686.

Dickinson, Terence *Hubble's Universe: Greatest Discoveries and Latest Images*, 2017, ISBN-9781770859975.

Ellerbroek, Lucas, *Planet Hunters: The search for extraterrestrial life,* 2017, ASIN-B073S986GV.

Glaser, Chaya *Uranus: Cold and Blue,* ISBN-1627245677.

Greene, Thomas P. *James Webb Space Telescope,* 2013, NASA Technical Reports Server, ISBN-978-1289058685.

Greenburg, Richard *Unmasking Europa: The Search for Life on Jupiter's Ocean Moon,* 2008, ISBN-978-0387479361.

Hand, Kevin *Alien Oceans: The Search for Life in the Depths of Space,* 2020, ISBN-978-0691179513.

Henin, Bernard *Exploring the Ocean Worlds of Our Solar System*, 2018, ISBN-978-3319934754.

Howell, Elizabeth "JUICE: Exploring Jupiter's Moons" 2017, Space.com.

Irwin, Patrick G. J. *Giant Planets of Our Solar System: Atmospheres, Composition, and Structure.* 2003 Springer, ISBN-3-540-00681-8.

"Lakdawalla, Emily "JUICE: Europe's next mission to Jupiter?" 2012, The Planetary Society.

Langley, Andrew, *Planet Hunting: Racking Up Data and Looking for Life,* 2019, ISBN-978-1543572704.

Ley, Willy *Gas Giants, the Largest Planets,* 1969, ISBN-978-0070376373.

Lorentz, Ralph, *Titan Unveiled: Saturn's Mysterious Moon Explored*, 2008, ISBN-978-0691125879

Lorentz, Ralph, *Saturn's Moon Titan: From 4.5 billion years ago to the present - An insight into the workings and exploration of the most Earth-like world in the outer solar system,* 2020, ISBN-978-1785216435.

Lowell, Percival, *Memoir On A Trans-neptunian Planet*, 2015, ISBN - 978-1296847647.

NASA, "NASA Astrophysics Missions: Reviews of Operating Missions - Hubble Space Telescope, Chandra X-ray Observatory, Fermi Gamma-ray Telescope, Kepler, Planck, Suzaku, Swift, Warm Spitzer, XMM-Newton." 2012, ASIN-B007SH9FCM.

Pappalardo, Robert T. (Ed), McKinnon, William B. (Ed), Khurana. Krishan (Ed), *Europa*, 2009, ISBN-978-0816528448.

Regius, Codex *Enceladus - Iceland of Space: The Cassini spacecraft over the moon of chilly geysers*, 2017, ASIN-B078R3J1VZ.

Schenk, Paul M., Clark, Roger N. *Enceladus and the Icy Moons of Saturn*, 2018, ISBN-978-0816537075.

Schenk, Paul M., Clark, Roger N. *Enceladus and the Icy Moons of Saturn,* 2018, ISBN-978-0816537075.

Sekbach, Joseph, Stan-Lotter Helga, *Extremophiles as Astrobiological Models,* 2020, ASIN B08QCPMFWB.

Stakem, Patrick H. *Robotic Exploration of the Icy Moons of the Gas Giants*, 2020, ISBN-979-8621431006.

Stern, S. Alan; Moore, Jeffrey M et al. The Pluto System After New Horizons, 2021, ISBN-978-0816540945.

Stakem, Patrick H., Martinez, Jose Carlos; Chandrarasenan, Vishnu; Mittra, Yash "A Cubesat Swarm Approach for Exploration of the Asteroid Belt," 2018, Presented to NASA Goddard Planetary CubeSats Symposium, 2018.

Stakem, Patrick H.; Da Costa, Rodrigo Santos Valente; Rezende, Aryadne; Ravazzi, Andre "A Cubesat-based alternative for the Juno Mission to Jupiter," 2017, Presentation to Flight Software Conference-17, JHU-APl

Stern, Alan *Chasing New Horizons: Inside the Epic First Mission to Pluto,* 2018, 978-1250098962.

Summers, Michael *Exoplanets: Diamond Worlds, Super Earths, Pulsar Planets, and the New Search for Life beyond Our Solar System,* 2017, Smithsonian Books, ASIN-B01HA426MS.

Resources

- www.planetary.org

- https://www.nasa.gov/specials/ocean-worlds/

- Exoplanet Travel Bureau, https://exoplanets.nasa.gov/alienworlds/exoplanet-travel-bureau/

- https://solarsystem.nasa.gov/moons/saturn-moons/titan/overview

- Nature, Life under the Ice, June 24, 2015, 522, 392

- Fox, Chris; Wiegert, Paul. "Exomoon Candidates from Transit Timing Variations: Eight Kepler systems with TTVs explainable by photometrically unseen exomoons" Monthly Notices of the Royal Astronomical Society, 23 November 2020.

- Kipping, David "An Independent Analysis of the Six Recently Claimed Exomoon Candidates". The Astrophysical Journal, 8 August 2020.

- Kipping, David; Bryson, Steve; et al."An exomoon survey of 70 cool giant exoplanets and the new candidate Kepler-1708 b-i". Nature, 13 Jan, 2022.

- "Astronomers may have found a huge moon around a Jupiter-like exoplanet". New Scientist. 28 January 2022.

- https://ssed.gsfc.nasa.gov/osean/science.html#papers

- https://www.lpi.usra.edu/opag/mar2012/presentations/Friday/5_JUICE_Summary.pdf

- https://sci.esa.int/web/juice

- https://www.lpi.usra.edu/opag/mar2012/presentations/Friday/5_JUICE_Summary.pdf

- https://meetingorganizer.copernicus.org/EPSC-DPS2011/EPSC-DPS2011-1343-1.pdf
-
- Joint NASA/ESA report on the TandEM/TSSM mission.

- "Complete Guide to NASA's James Webb Space Telescope (JWST) Project - Spacecraft, Instruments and Mirror, Science, Infrared Astronomy, GAO and Independent Review Reports, Congressional Hearings, 2011, ISBN-1549878212.

- Wikipedia, various.

If you enjoyed this book, you might be interested in some of his others as well.

Stakem, Patrick H. *Floating Point Computation*, 2013, PRRB Publishing, ISBN-152021619X.

Stakem, Patrick H. *Architecture of Massively Parallel Microprocessor Systems*, 2011, PRRB Publishing, ISBN-1520250061.

Stakem, Patrick H. *Multicore Computer Architecture,* 2014, PRRB Publishing, ISBN-1520241372.

Stakem, Patrick H. *Personal Robots*, 2014, PRRB Publishing, ISBN-1520216254.

Stakem, Patrick H. *RISC Microprocessors, History and Overview,* 2013, PRRB Publishing, ISBN-1520216289.

Stakem, Patrick H. *Robots and Telerobots in Space Application*s, 2011, PRRB Publishing, ISBN-1520210361.

Stakem, Patrick H. *The Saturn Rocket and the Pegasus Missions, 1965,* 2013, PRRB Publishing, ISBN-1520209916.

Stakem, Patrick H. *Visiting the NASA Centers, and Locations of Historic Rockets & Spacecraft,* 2017, PRRB Publishing, ISBN-1549651205.

Stakem, Patrick H. *Microprocessors in Space*, 2011, PRRB Publishing, ISBN-1520216343.

Stakem, Patrick H. Computer *Virtualization and the Cloud*, 2013, PRRB Publishing, ISBN-152021636X.

Stakem, Patrick H. *What's the Worst That Could Happen? Bad Assumptions, Ignorance, Failures and Screw-ups in Engineering Projects,* 2014, PRRB Publishing, ISBN-1520207166.

Stakem, Patrick H. *Computer Architecture & Programming of the Intel x86 Family,* 2013, PRRB Publishing, ISBN-1520263724.

Stakem, Patrick H. *The Hardware and Software Architecture of the Transputer*, 2011,PRRB Publishing, ISBN-152020681X.

Stakem, Patrick H. *Mainframes, Computing on Big Iron*, 2015, PRRB Publishing, ISBN- 1520216459.

Stakem, Patrick H. *Spacecraft Control Centers*, 2015, PRRB Publishing, ISBN-1520200617.

Stakem, Patrick H. *Embedded in Space,* 2015, PRRB Publishing, ISBN-1520215916.

Stakem, Patrick H. *A Practitioner's Guide to RISC Microprocessor Architecture*, Wiley-Interscience, 1996, ISBN-0471130184.

Stakem, Patrick H. *Cubesat Engineering*, PRRB Publishing, 2017, ISBN-1520754019.

Stakem, Patrick H. *Cubesat Operations*, PRRB Publishing, 2017, ISBN-152076717X.

Stakem, Patrick H. *Interplanetary Cubesats*, PRRB Publishing, 2017, ISBN-1520766173 .

Stakem, Patrick H. Cubesat Constellations, Clusters, and Swarms, Stakem, PRRB Publishing, 2017, ISBN-1520767544.

Stakem, Patrick H. *Graphics Processing Units, an overview*, 2017, PRRB Publishing, ISBN-1520879695.

Stakem, Patrick H. *Intel Embedded and the Arduino-101, 2017,* PRRB Publishing, ISBN-1520879296.

Stakem, Patrick H. *Orbital Debris, the problem and the mitigation*, 2018, PRRB Publishing, ISBN-*1980466483*.

Stakem, Patrick H. *Manufacturing in Space*, 2018, PRRB Publishing, ISBN-1977076041.

Stakem, Patrick H. *NASA's Ships and Planes*, 2018, PRRB Publishing, ISBN-1977076823.

Stakem, Patrick H. *Space Tourism*, 2018, PRRB Publishing, ISBN-1977073506.

Stakem, Patrick H. *STEM – Data Storage and Communications*, 2018, PRRB Publishing, ISBN-1977073115.

Stakem, Patrick H. *In-Space Robotic Repair and Servicing*, 2018, PRRB Publishing, ISBN-1980478236.

Stakem, Patrick H. *Introducing Weather in the pre-K to 12 Curricula, A Resource Guide for Educators*, 2017, PRRB Publishing, ISBN-1980638241.

Stakem, Patrick H. *Introducing Astronomy in the pre-K to 12 Curricula, A Resource Guide for Educators*, 2017, PRRB Publishing, ISBN-198104065X.
Also available in a Brazilian Portuguese edition, ISBN-1983106127.

Stakem, Patrick H. *Deep Space Gateways, the Moon and Beyond*, 2017, PRRB Publishing, ISBN-1973465701.

Stakem, Patrick H. *Exploration of the Gas Giants, Space Missions to Jupiter, Saturn, Uranus, and Neptune*, PRRB Publishing, 2018, ISBN-9781717814500.

Stakem, Patrick H. *Crewed Spacecraft*, 2017, PRRB Publishing, ISBN-1549992406.

Stakem, Patrick H. *Rocketplanes to Space*, 2017, PRRB Publishing, ISBN-1549992589.

Stakem, Patrick H. *Crewed Space Stations,* 2017, PRRB Publishing, ISBN-1549992228.

Stakem, Patrick H. *Enviro-bots for STEM: Using Robotics in the pre-K to 12 Curricula, A Resource Guide for Educators,* 2017, PRRB Publishing, ISBN-1549656619.

Stakem, Patrick H. *STEM-Sat, Using Cubesats in the pre-K to 12 Curricula, A Resource Guide for Educators*, 2017, ISBN-1549656376.

Stakem, Patrick H. *Lunar Orbital Platform-Gateway*, 2018, PRRB Publishing, ISBN-1980498628.

Stakem, Patrick H. *Embedded GPU's*, 2018, PRRB Publishing, ISBN- 1980476497.

Stakem, Patrick H. *Mobile Cloud Robotics*, 2018, PRRB Publishing, ISBN- 1980488088.

Stakem, Patrick H. *Extreme Environment Embedded Systems,* 2017, PRRB Publishing, ISBN-1520215967.

Stakem, Patrick H. *What's the Worst, Volume-2*, 2018, ISBN-1981005579.

Stakem, Patrick H., *Spaceports*, 2018, ISBN-1981022287.

Stakem, Patrick H., *Space Launch Vehicles*, 2018, ISBN-1983071773.

Stakem, Patrick H. *Mars*, 2018, ISBN-1983116902.

Stakem, Patrick H. *X-86, 40th Anniversary ed*, 2018, ISBN-1983189405.

Stakem, Patrick H. *Lunar Orbital Platform-Gateway*, 2018, PRRB Publishing, ISBN-1980498628.

Stakem, Patrick H. *Space Weather*, 2018, ISBN-1723904023.

Stakem, Patrick H. *STEM-Engineering Process*, 2017, ISBN-1983196517.

Stakem, Patrick H. *Space Telescopes,* 2018, PRRB Publishing, ISBN-1728728568.

Stakem, Patrick H. *Exoplanets*, 2018, PRRB Publishing, ISBN-9781731385055.

Stakem, Patrick H. *Planetary Defense*, 2018, PRRB Publishing, ISBN-9781731001207.

Patrick H. Stakem *Exploration of the Asteroid Belt*, 2018, PRRB Publishing, ISBN-1731049846.

Patrick H. Stakem *Terraforming*, 2018, PRRB Publishing, ISBN-1790308100.

Patrick H. Stakem, *Martian Railroad,* 2019, PRRB Publishing, ISBN-1794488243.

Patrick H. Stakem, *Exoplanets,* 2019, PRRB Publishing, ISBN-1731385056.

Patrick H. Stakem, *Exploiting the Moon,* 2019, PRRB Publishing, ISBN-1091057850.

Patrick H. Stakem, *RISC-V, an Open Source Solution for Space Flight Computers,* 2019, PRRB Publishing, ISBN-1796434388.

Patrick H. Stakem, *Arm in Space*, 2019, PRRB Publishing, ISBN-9781099789137.

Patrick H. Stakem, *Extraterrestrial Life*, 2019, PRRB Publishing, ISBN-978-1072072188.

Patrick H. Stakem, *Space Command*, 2019, PRRB Publishing, ISBN-978-1693005398.

CubeRovers, A Synergy of Technologys, 2020, PRRB Publishing, ISBN-979-8651773138.

Robotic Exploration of the Icy moons of the Gas Giants. 2020, PRRB Publishing, ISBN- 979-8621431006

Hacking Cubesats, 2020, PRRB Publishing, ISBN-979-8623458964.

History & Future of Cubesats, PRRB Publishing, ISBN-979-8649179386.

Hacking Cubesats, Cybersecurity in Space, 2020, PRRB Publishing, ISBN-979-8623458964.

Powerships, Powerbarges, Floating Wind Farms: electricity when and where you need it, 2021, PRRB Publishing, ISBN-979-8716199477.

Hospital Ships, Trains, and Aircraft, 2020, PRRB Publishing, ISBN-979-8642944349.

CubeRovers, a Synergy of Technologys, 2020, ISBN-979-8651773138

Exploration of Lunar & Martian Lava Tubes by Cube-X, ISBN-979-8621435325.

Robotic Exploration of the Icy moons of the Gas Giants, ISBN- 979-8621431006.

History & Future of Cubesats, ISBN-978-1986536356.

Robotic Exploration of the Icy Moons of the Ice Giants, by Swarms of Cubesats, ISBN-979-8621431006.

Swarm Robotics, ISBN-979-8534505948.

Introduction to Electric Power Systems, ISBN-979-8519208727.

Centros de Control: Operaciones en Satélites del Estándar CubeSat (Spanish Edition), 2021, ISBN-979-8510113068.

Exploration of Venus, 2022, ISBN-979-8484416110.

Patrick H. Stakem, *The Search for Extraterrestial Life,* 2019, PRRB Publishing, ISBN-1072072181.

The Artemis Missions, Return to the Moon, and on to Mars, 2021, ISBN-979-8490532361.

James Webb Space Telescope. A New Era in Astronomy, 2021, ISBN-979-8773857969.